STECK-VAUGHN
PAIR-IT BOOKS

D0074378

I'm a Little Seed

Written by Kay Sands
Illustrated by Benrei Huang

STECK-VAUGHN
ELEMENTARY • SECONDARY • ADULT • LIBRARY
A Harcourt Classroom Education Company

www.steck-vaughn.com

I'm a little seed,

short and stout.

Up through the ground,

I grow a sprout.

Then I grow a stem,

and some green leaves.

Someday I will be a tree.